EG Practice
The surveyor as expert witness:
Building and development play

A Bruce Shaw play
by Alan Morris

Estates Gazette, 1 Procter Street
London WC1V 6EU
A division of Reed Business Information

EG Books

Commissioning Editor: Alison Richards
Typesetting: Isabel Micallef — in Meta and Palatino
Cover and interior design: Deborah Ford, EG
Printed and bound by: Hobbs the Printers Limited, Totton, Hampshire

Contents

For my wife Angela

Foreword

I am pleased to write a few words about this play. The purpose of the play is to promote a better understanding of the role of expert witness — in such a manner as to engage the enquirer. In these times, the expert witness is much in the news: it is of key importance that the role be understood, and properly discharged. It is a role that is by no means easy to discharge, and even the seasoned professional can benefit from appropriate training.

Why through the medium of a play? A play of this kind serves several uses: it can teach; it can help learning; and it can be quite memorable. Above all, it can be enjoyable (and not a little amusing) both to the audience and to the players.

I am therefore delighted to play a part in launching this play.

Felix Collins, Managing Partner at Bruce Shaw Partnership

Preface

I wrote this play some time ago, for a meeting of the Royal Institution of Chartered Surveyors. The purpose of the play is to give an explanation, in an entertaining way, of the role of the expert witness; and also to provide tips for those engaged in expert witness work. The play comprises a sequence of short scenes, with commentary between, tracing the fortunes of an expert witness in lively action. The play is flexible as to the proceedings — whether litigation, arbitration or adjudication proceedings.

The role of expert witness makes considerable demands upon anyone who ventures to cross its threshold: for he will have a duty of allegiance to the court or tribunal while employed and paid by just one of the parties to the action.

The party who pays can, in all reasonableness, look for service from the expert. Yet the expert, in the matter of his opinion, must give allegiance to the tribunal. He must stand aloof to partisan opinion and with true opinion assist the tribunal towards the achievement of a just decision.

In the course of the proceedings, the expert will be faced with cross-examination: that is, focused questioning as to whether he really is an expert and then as to the soundness of his opinion. The expert must be careful not to fall foul of counsel's wiles; not find himself agreeing with counsel's propositions where he, himself, would not have put those propositions in quite the same way. The play brings out all of these considerations in the context of a lively portrayal.

The play has been performed at a variety of events: at meetings of professional bodies — the Royal Institution of Chartered Surveyors, the Chartered Institute of Arbitrators; and also at events at universities and firms. A play-reading basis has been the norm at these events, with spirited participants ready to hand to take up the parts: sometimes participants direct from the audience, sometimes the organisers of the event. Where the event is to be on a play-reading basis, it is recommended that those taking up the parts read the play prior to the event.

Whatever the occasion — be it a period of private study or a public performance — the play will provide an enjoyable, informative and educational experience.

Depending upon the time available, it may be preferred to omit part of the text. In that case, *Scene 3: Meeting of the experts* can be omitted without losing too much of the continuity of the play.

In recent years, a great deal of thought has been given to the conduct required of the expert witness in court, arbitration and other judicial proceedings. Practice statements and guidance notes have been issued. I have in mind here: the *Code of Guidance for Experts under the Civil Procedure Rules*, the *Technology and Construction Solicitors Association Expert Witness Protocol*; also, the publication *Surveyors Acting as Expert Witnesses* (the practice statement and guidance notes of the Royal Institution of Chartered Surveyors). The advice given in these publications is held very much in mind in the text of the play.

I am grateful to Estates Gazette Books and Bruce Shaw Partnership for their ready support and encouragement towards the publication of the play; also to my colleague John Pointing, for reading through the play and making suggestions for improvement.

Alan Morris

Dramatis personae

Commentator

Solicitor

Claimant's expert witness

Respondent's expert witness

Arbitrator or judge or adjudicator

Counsel for the claimant (examining counsel)

Counsel for the respondent (cross-examining counsel)

Introductory commentary

Commentator: The purpose of this play and commentary is to give something of the position of the expert witness in court, arbitration or adjudication proceedings.

There is, in these proceedings, a place for two kinds of witness: the *witness of fact*; and the *expert witness*.

Witness of fact: here the witness comes to the proceedings with knowledge of some fact or facts relating to the dispute (for example what someone did on a particular occasion) and gives evidence of those facts.

Expert witness: this witness comes to the proceedings on account of his[1] expertise enabling him to make *deductions* from the facts. He will be able to advise, for example, whether the particular piece of work should, or should not, rightly be reckoned below the normal standard for the profession, such as to constitute *professional negligence*. He does not normally come to the proceedings with knowledge of the facts. It is his first task to search out or otherwise become informed of the facts.

So we distinguish between the *witness of fact* and the *expert witness*: that the former gives factual evidence; whereas the latter makes deductions from factual evidence. The expert witness assists the court in its task of *deciding*: whether the evidence leads — or does not lead — to what is alleged.

Although the expert's function is to assist the court, he is normally appointed by one of the parties. Usually, each party has its own expert witness — whom he pays. This may be a matter of surprise — where the expert is to assist the court and is not to be advocate for the party.

This position — that each party has its own expert witness — goes back to the *adversarial* system prevalent in jurisdictions in which the administration of justice calls upon the *common law*. Here, the expert is cross-examined by counsel for the opposing party. This is with a view to discrediting the witness. The expert must be sure to prepare himself carefully — as we shall see.[2]

Now for the play.

1. Throughout the play, wherever he is mentioned it is inclusive of she and wherever his is mentioned it is inclusive of her.
2. It is significant, notwithstanding the above, that more recent provision has introduced power for the tribunal to call its own expert witness in certain circumstances. Perhaps this is indicative of a softening of the adversarial stance in some quarters.

Solicitor: Things are certainly different from what I thought. Let us take stock from the beginning.

(pause, then slowly and emphatically) There are *three* elements here I suggest: *first,* there is a low initial estimate — that the surveyor, later, tried to cover up; and *second,* we have the additional items (I must hear from you how much you reckon for these items); and *third,* there is excessive labour ... as you say.

(slowly, thoughtful) And the final account is for £900,000?

Expert witness: Yes, that's right.

Solicitor: That's £400,000 more than the initial estimate.

Now what proportion of that £400,000 do you put to each of the three elements: the low estimate; the client's added items; the excessive cost for labour?

First the low estimate.

Expert witness: *(slowly, referring to a piece of paper)* Well, £180,000 goes to the low estimate. Instead of £500,000, the estimate at the beginning should have been £680,000.

Solicitor: And what about the client's added items: what do you reckon is the proper amount for them?

Expert witness: I reckon they should be about £160,000. This means that the figure of £400,000 for these items in the final account is about £240,000 too much.

Solicitor: And now what about the excessive labour?

Expert witness: I reckon that to be about £60,000.

Solicitor: (*totalling up*) Yes, £400,000 in all.

Expert witness: That's right.

All in all, the final account figure should be, in my estimation, £840,000: that is to say, £60,000 less than the £900,000 paid.[3]

Solicitor: *(reflective)* I see. So we have, at the beginning, the surveyor's initial estimate prepared without due care and skill; and then, at the end, in the final account, excessive labour costs approved.

And we have payment made in excess of what it should be, but not to the extent that we might have thought.

Expert witness: Correct.

Solicitor: *(pause, then slowly)* What you have just told me makes things difficult for recovering money. The surveyor's low estimate may not help us, as I say.

Expert witness: Why is that?

Solicitor: The trouble is, there are two things to show in order to be entitled in a claim against the surveyor. First, there has to be *liability;* and then there has to be *loss* incurred.

3. It might be thought worthwhile to provide the audience with these figures in hard copy. A suggestion for the format of this is provided on page 29.

I think we have breach of contract on the part of the surveyor giving rise to *liability*; but then, strictly speaking, no *loss* — no loss resulting from the low estimate.

Expert witness: *(taking stock)* I suppose you are right.

Solicitor: *(decisively)* It's like this, you see. The client undertook to pay the builder's cost. To that extent he has suffered no loss. The estimate is neither here nor there as regards that point.

(slight pause)

Expert witness: But then *(hesitates)* that hardly seems right? The value of the house is mostly unaltered with the work, I expect — it is much the same now as it was before. So the client has lost a very great deal: all of his expenditure of £900,000.

Solicitor: *(short laugh)* I am afraid that won't do. My client hasn't the sale of his house in mind. He plans to live there for the foreseeable future. He has followed his own whims, without any thought about market value.

The dining room is turned into an Aladdin's cave. There is a lamp on the table. If ever you rub it, sparks come down the chimney; and then a voice booms out *(booming)* "What do you want for your dessert, oh master?"

Not everyone looks for *that* sort of thing when they buy a house.

As you say, the expense has not enhanced the market value of the house. But the law does not reckon that to be a loss. Oh no.

Expert witness: *(rather crestfallen)* I see. So there is no future in that line of approach?

Solicitor: Very sorry.

Expert witness: *(rather timidly)* And there was another aspect that I thought to mention.

Solicitor: Well, what's that?

Expert witness: Would your client have gone ahead at all if, at the beginning, he had known the true cost? If properly advised he might well have been saved all his expense — by not going ahead.

Solicitor: I have been thinking about that. But then we have no evidence by way of proof that he would *not* have gone ahead. In fact, as I know him, I think he would have done.

(some finality) So, I suspect there is no future in that argument, either.

No, we must look elsewhere than the surveyor's low estimate if we are going to achieve much success for my client.

(mildly questioning) The loss, it seems, is £60,000 — and this because, as you say, the surveyor allowed through too much labour.

Expert witness: Yes, that's right.

Solicitor: *(after a slight pause)* Ah yes! Let's have some lunch. Tell me about your figures while we eat. How is it that they show there's too much labour in the final account? Just how *sure* are you?

Commentator: Off they go to lunch, so that the expert witness can explain to the solicitor the basis on which he has come to his conclusion that there is too much labour in the surveyor's account.

During this short interlude, let us catch up with the "other side" — the surveyor's position. What of that side? The surveyor's solicitor will be engaging in much the same sort of thing as we have seen on the client's side. He, *also*, will have brought in an expert. The expert will have written *his* report. The solicitor will be taking stock of the position.

So, what now?

Will the parties settle? They each, at this stage, have their expert's opinion to help them in appraising the position they are holding. Will that help?

We must presume (in the interest of continuing our play!) that the parties do not settle. Probably the experts hold different views: such that both sides reckon there is sufficient in their case for continuing with the action.

So ... directions will be issued for the experts to meet. The solicitors will agree the agenda for the meeting. Who knows whether, in quiet discussion at the meeting, one of the experts may not be persuaded to the other's viewpoint?

Meeting of the experts

3

Commentator: Scene 3 Meeting of the experts. The experts meet, as directed. Now, just why a meeting of the experts? It is to identify any differences in their opinions and attempt agreement of some of the issues — maybe them all.

But first there is a procedural point touching the exchanging of the experts' reports. Should the exchange precede their meeting? Or should their meeting precede the exchange?

In our play, we have the meeting preceding the exchange — the procedure we prefer. To exchange reports before the meeting would be at this risk: that differences would be hardened, rendering difficult any changing of viewpoint.

Now — to the meeting.

Respondent's expert witness: *(voluble throughout)* Good morning. I am John Vintner. You have had a look at things on behalf of the claimant?

Claimant's expert witness: Yes, that's right.

Respondent's expert witness: Good. I have done the same — for the respondent.

Everything without prejudice here today, of course.

We will have to find out if we are of one mind.

Claimant's expert witness: Yes, of course. I hope we are.

Respondent's expert witness: Now then ... where's that agenda? *(pulls out a piece of paper)* Ah, yes. *Item 1 — initial remarks.* Well, since you are for the claimant, I will stand back. Will you start?

Claimant's expert witness: *(rather emphatic)* Yes, certainly I will. My view is that the additional items (you know what they are of course? — the work beyond what was in the specification) should not have cost anything like as much as the surveyor has said in his final account.

Respondent's expert witness: Yes. I have come to that conclusion too.

Claimant's expert witness: *(relieved)* Really? That's a good start.

Respondent's expert witness: Why, yes it is. It isn't often that there is agreement right from the start.

It is like this, you see. They asked me to look over the surveyor's figures. The surveyor made over to me all his calculations. It was quite clear, from them, just what had happened. The surveyor's estimate at the start of the job was much too low.

There were, in essence, these three figures:

- £500,000 — the surveyor's estimate at the start of the job, including the fixed fee.
- £400,000 — an amount for the client's added items.
- £900,000 — the total amount of the final account.

Claimant's expert witness: That is clear enough, but ...

Respondent's expert witness: *(carrying on regardless — rather ponderous)* Now about these three figures ...

The total amount — £900,000: no fault there — no doubt about that.

The other two figures — original estimate £500,000; additional items £400,000 — here there was, quite clearly, something wrong. There was too much in the additional items and too little in the surveyor's estimate to cover, as it should, the original works.

How this came about is a sorry story: although, perhaps, understandable. It's a nasty example of slipshod work. I cannot abide that sort of thing. (I can be perfectly frank, since we are *without prejudice*).

It happened like this. At final account, the surveyor arrived at the total cost of £900,000. He then simply deducted the amount of his estimate — £500,000 — and he put the rest to the additional items — £400,000. Just like that. No checking at all.

All he gave thought to was the total figure. He was confident of the total and went no further.

(ruminating) I suppose, in a sense, the figure that really matters *is* the total amount. *(petulant)* But things like this cause such unnecessary aggravation, don't you agree?

Claimant's expert witness: This is all very well, but ...

Respondent's expert witness: *(charging on regardless)* The first thing the surveyor knew, was a telephone call from his client in a decidedly upset frame of mind. He was questioning the cost for the additional items in the final account. Knowing what we know, this was perfectly understandable.

But it took the surveyor completely by surprise. He had focused his attention on the total figure. The total seemed the important thing. It was only as an afterthought that he had split the total figure in the final account.

Claimant's expert witness: This is all very well, but ...

Respondent's expert witness: *(rather firm)* Just hear me out, please. Now, as it happened, he was busy at the time of receiving that telephone call from the client; and so it slipped his memory to look into the matter. Then, suddenly, as he says, he had the client claiming — all the rest of it.

The surveyor seems not ever to have questioned whether his estimate had been wrong. *(darkly)* Although to my mind (always *without prejudice*, of course), he might have *thought* but not wanted to *say*. He had given firm assurances about his estimate earlier on.

Well, then, in all modesty, *I (pointing to himself)* was appointed. I saw, within five minutes, where the trouble lay. There was too little in that original estimate. Would you believe it, in collecting up the figures, he had missed out half the pages.

I spoke with the solicitor ... *(deliberately, tone of finality)* and we agreed that I should tell you about it at our meeting here today *(beaming at the claimant's expert witness)*, which I have now done. It explains the whole thing.

Claimant's expert witness: *(rather sternly)* That is all very well, but you are overlooking something.

Respondent's expert witness: *(very surprised)* Why? What's that?

Claimant's expert witness: You have given one reason — but there are two reasons, you know, why the amount for the additional items is much higher than it should be.

Respondent's expert witness: Well I never! You do surprise me.

Claimant's expert witness: There was an error, I grant you, in the initial estimate. I agree with you there. But, quite apart from the initial estimate, the *total figure*'s wrong: £900,000 is too much, you know.

Respondent's expert witness: This is not something that I have found at all.

Claimant's expert witness: But yes, it is so.
The *labour content* in the final account is more than it should be, in my submission. And, from that, it follows that the final account figure is higher than it should be.
This is something *quite apart from the initial estimate.*

Respondent's expert witness: *(taken aback)* I can not agree.
(pointing-out tone) There are time sheets for all the labour, of course?

Claimant's expert witness: Oh yes, there are.

Respondent's expert witness: *(continuing uninterrupted — rather forceful)* Unless you are suggesting that some of the labour was booked dishonestly?
(firmly) I could not agree there was anything of *that* sort.

Claimant's expert witness: I'm not here to make any suggestion — one way or the other. I simply know that I have analysed the job, and find too much labour in the final account.

Respondent's expert witness: Well, this is something I *hadn't* expected.
(challenging) Come to think of it, you've quite a job on your hands to show that there's too much labour, haven't you? Just how *much* too much labour are you saying?

Claimant's expert witness: In money terms — £60,000.

Respondent's expert witness: *(thoughtful)* I see. Let's now think.
(musing to *himself — slowly)* £60,000. Now, what proportion of the total labour does that work out at? *(pause)* About 15%?
(then continues briskly) Well, that is up to you of course — about the labour. I had expected my explanation to see the end of this dispute.
(concluding tone) So ... Perhaps that is as far as we can go at present?

Claimant's expert witness: That seems to be the case.

Respondent's expert witness: Let me sum up. We still have, outstanding, the amount of labour. But we are of one mind about the initial estimate: that the error in the estimate had no bearing on what was to be paid at the final account.

Claimant's expert witness: Yes, that's quite right.

Respondent's expert witness: Just to be quite clear about my position: this dispute is about the surveyor's slipshod work. That is regretted. But what we have in the final account is the proper price, according to the contract, for the work done.

(firmly) I do hope that, upon reflection, you will persuade your side to drop the case: that there really is no point in pursuing the matter any further.

Claimant's expert witness: I am sure we both hoped we could reach complete agreement. But, unfortunately, not.

What next, now? Ah, yes. We must go to our *statement of agreement and disagreement* — do that now.

Then on to writing our formal reports. We should soon be receiving a direction for exchange.

(they set about their statement)

Commentator: Note that the experts, at the end of their discussion, prepare a statement, which they both sign, of the areas of agreement and disagreement.

Note also, that the respondent's expert witness, for all his blustering and verbosity, seems to know his job. Did you notice how he picked up the disputed labour as a proportion of the total labour cost? We will find ourselves returning to this, later on.

Commentator: Scene 4 Examination, cross-examination. So we come to the hearing: though it is by no means always that there will be a hearing. It may be that the award will be made on the basis of *documents only*: that the judgement or award will be decided on the basis of a reading of the documents, without there being recourse to a hearing.

There is much to be said for the *documents only* basis. The arrangements for a hearing will always take time and the hearing itself will be expensive and lengthy.

So why have a hearing? It is primarily in order to test the evidence given in the documents, through face-to-face questioning.

We enter upon the hearing at the point at which the expert witness is called — the claimant's expert witness, whose fortunes we are following. The expert witness is questioned, at first, by friendly counsel — counsel for the claimant. Then follows *cross-examination* — probing questioning by counsel for the respondent. Cross-examining counsel will probe the disagreement that the experts recorded at the end of their discussion.

Examination, cross-examination

4

Examining counsel: *(announcing tone)* I call expert witness for the claimant. *(pause for expert witness to come forward)*

Are you Arthur Simpson, expert witness for the claimant?

Claimant's expert witness: Yes.

Examining counsel: *(to expert witness)* Do you practise under the style Earnest Bright & Partners, Surveyors of 10 Bracken Street, Leamington Spa?

Expert witness: I do.

Examining counsel: Will you outline your experience as a practising surveyor?

Expert witness: I qualified as a surveyor in 1975 and have practised as a surveyor engaged upon building contracts from that time. In 1984, I became senior partner of my firm Earnest Bright and have remained in that capacity ever since. Over the past 20 years, I have served as an expert witness on many occasions.

Examining counsel: Have you acquainted yourself with the present dispute between the parties?

Expert witness: Yes, I have.

Examining counsel: Have you met with the respondent's expert — on a *without prejudice* basis — with a view to achieving agreement leading to a settlement of the dispute?

Expert witness: Yes, I have.

Examining counsel: At the end of your discussion with the respondent's expert did you prepare and agree a statement of the areas of agreement and disagreement?

Expert witness: Yes, I did

Examining counsel: Have you subsequently exchanged reports with the respondent's expert?

Expert witness: Yes, I have.

Examining counsel: Is this your report? *(hands document to expert)*

Expert witness: *(examines the document)* Yes.

Examining counsel: Will you please turn to page 32 of the report. *(expert witness opens the document)* Can you confirm that that is your signature at the end of the report?

Expert witness: Yes, it is.

Examining counsel: Is there any part of the report that you wish to withdraw or any addition that you wish to make?

Expert witness: No.

(slight pause)

Examining counsel: *(new subject)* Have you seen this document before? *(hands final account to the expert witness).*

Expert witness: *(examines the document)* Yes. It is the final account for the refurbishment works at 39 *The Oaks, Nuneaton* — the works that are the subject-matter of the dispute.

Examining counsel: What observations have you on the final account?

Expert witness: *(gets out some sheets of paper and addresses arbitrator[4])* I would like to refer to my notes with your permission, Sir.

Arbitrator:[4] Yes, you have my permission.

Examining counsel: *(to the arbitrator)* I have a copy of the notes for you here, Sir.[5]

(hands sheets to the arbitrator)

Arbitrator: Thank you.

Examining counsel: *(again to the arbitrator)* Respondent's counsel already has a copy of the notes.

Arbitrator: *(to counsel for the respondent)* That is so?

Counsel for the respondent: *(stands)* Yes, Sir.

Arbitrator: *(looks at sheets)*
(then to examining counsel) Please proceed.

Examining counsel: *(repeats question to expert witness)* What observations have you on the final account?

4. Throughout this scene, the term "arbitrator" is adopted as inclusive of "judge" or "adjudicator".
5. For proceedings before an arbitrator, the term "Sir" is used. For proceedings in the county court before a District Judge or Recorder, the term "Your Honour" would be used. For proceedings before an adjudicator, the term "Sir" would be used.

Expert witness: The document seems to me not very informative.

Examining counsel: What does it say?

Expert witness: It starts with the surveyor's estimate for the works, then lists five additional items.

It states:

Final account for works carried out at 39 The Oaks, Nuneaton

All as estimate of 23 September 2004 — £500,000

Additional items:

- Renew front door
- Take down plaster and renew — kitchen
- Bricking up window opening — study
- Renewal of felt roofing — porch
- Redecoration — dining room.

Total additional items	£400,000
Total	£900,000.

Examining counsel: And what is your opinion of the reasonableness of this account?

Expert witness: The account does not appear at all reasonable to me.

Examining counsel: In what way is that?

Expert witness: Well, in the first place, the overall total is higher than can be justified. That is the first thing to say about the account.

Beyond that, there is the manner of presentation of the account. It does not reflect what is in the contract at all.

The amount for the original works is given in terms of the surveyor's estimate — where the *actual amount under the contract* should be given — on the basis of *cost* and *builder's fee*.

Again, the amount for the "additional items", as stated, is exorbitant.

Altogether, the final account should give much more information about the basis for the figures.

Examining counsel: You say the amount for the additional items is exorbitant. What reason have you for that opinion?

Expert witness: Well, I have priced the additional items and they come to much less than the surveyor has for them. It has to be said, though, that *the amount in the account for the original work is much less than it should be.* To a certain extent, this cancels out the excessive amount for the additional work.

Examining counsel: But taking all of this into account, you are of the opinion that the overall total is too high?

Expert witness: Yes

Examining counsel: Just why is it that the overall total is too high?

Expert witness: It is because too much labour has been accounted to the job.

Examining counsel: Is this just to do with the additional items?

Expert witness: No. It applies generally. It is on account of there not having been proper control exercised over the labour records throughout the job.

Examining counsel: What does the expert for the respondent say in his report about the labour?

Expert witness: He says that he does not find evidence that too much labour has been included.

Examining counsel: And to your mind, what you are saying is, in fact, the case: that there *is* too much labour?

Expert witness: Yes. And when I say there was no control exercised, it is to point out that the time sheets for the work were never signed or even checked. Yet all of the time recorded on the time sheets was included in the final account. That is not a good sign, you know.

Examining counsel: And you say that the additional items figure is high quite apart from question of the amount of labour?

Expert witness: Yes. It is high, also, because part of the cost of the original work has been attributed to the additional items. The respondent's expert agrees with me on this point.

Examining counsel: Does he agree as to the amount by which the figure for the additional items is high?

Expert witness: No, he doesn't. This is because he does not agree my point about the labour.

Examining counsel: So have you arrived at a price for the labour?

Expert witness: Yes, I have.

Examining counsel: How have you done that?

Expert witness: I have costed the work, then taken the labour as a proportion of that cost.

Examining counsel: And what figure have you arrived at for the additional items?

Expert witness: £160,000.

Examining counsel: So you have come to the conclusion that the final account figure for the additional items is over twice as much as it should be?

Expert witness: Yes.

Examining counsel: *(pause, new subject tone)* So, I understand you to have said there are two reasons for the difference between your figure and the surveyor's.

One of these reasons is the surveyor's low estimate — you have agreed this with the respondent's expert; and the other reason is the excessive labour — which the respondent's expert does not agree.

Expert witness: Yes.

Examining counsel: *("let us summarise" tone)* So how do you see the final account as it should be?

Expert witness: In my opinion, it should total £840,000. This figure is made up in the following way.

(very slowly) Taking the original works first, the amount should be £680,000 inclusive of fixed fee (this in place of £500,000 in the final account).

(pause) Now, the additional items — these should be £160,000 (this in place of £400,000 in the final account). *(pause)* Totalling-up the figures — £680,000 and £160,000 — brings us to £840,000.

Now the final account totals £900,000. So I see an excess of £60,000.

Examining counsel: Where did the surveyor get his information regarding the labour?

Expert witness: There were labour records for the job.

Examining counsel: Then how can you say that too much labour was allowed through in the final account?

Expert witness: It is my view that the time sheets submitted by the contractor should not have been accepted without question by the surveyor.

Examining counsel: Why is that?

Expert witness: Well, none of them were signed. It is normal to have time sheets signed by a representative of the employer, say a clerk of works, or someone like that. In our case, the records were not signed. Yet the surveyor accepted them, every bit.

Examining counsel: So how would you expect the records to be handled?

Expert witness: I would expect to see a process for verification put in place, with someone authorised to verify the records.

Examining counsel: Thank you.

(to the arbitrator) Sir, that concludes my examination of the witness,

Arbitrator: *(to examining counsel)* Thank you Mr Counsel. *(pause)*

(to cross-examining counsel) I call on counsel for the respondent to cross-examine the witness.

Cross-examining counsel: *(comes forward)* Thank you, Sir. *(pause)*

(to expert witness) You say you have been practising your profession of surveyor for 30 years?

Expert witness: Yes, 30 years.

Cross-examining counsel: And you qualified before that?

Expert witness: No, I studied by correspondence course in the evenings for five years while I was working in a firm during the daytime.

Cross-examining counsel: And have you kept up with your studies at all since then?

Expert witness: Yes. I have learned through experience at work.

Cross-examining counsel: *(very slightly threatening)* I asked you if you had kept up with your *studies* since then?

Expert witness: Well. ... I have attended short courses. You know, all surveyors are required to attend courses now.

Cross-examining counsel: *(slightly more threatening)* I was asking about you, not surveyors in general.
Now I want you to answer this question: how many courses have you been to this year?

Expert witness: *(nervous laugh)* I cannot be sure.

Cross-examining counsel: Let me help you. Have you been to any courses on *estimating* in the past year?

Expert witness: I cannot say that I have. Estimating is a practical skill; it is best learned on the job.

(pause)

Cross-examining counsel: How accurately can you estimate, do you think?

Expert witness: Quite well. Once or twice I have been spot on. One can't hope for that every time, of course.

Cross-examining counsel: Nothing more precise?

Expert witness: Oh, I can estimate within about 3%.

Cross-examining counsel: As close as that?

Expert witness: Well, no more than 5%, anyway.

Cross-examining counsel: 5% is almost twice as much as 3% isn't it?

Expert witness: Well ... yes.

Cross-examining counsel: Do you keep records of your performance? ... see how you are improving?

Expert witness: No, I cannot say that I do.

Counsel for the claimant: *(jumps up)* Sir, these questions are irrelevant to the proceedings before you. I raise an objection to the questions.

(pregnant silence)

Arbitrator: No, Mr Counsel, I do not accept your objection. The questions are directly relevant.

(to cross-examining counsel) Please proceed.

Cross-examining counsel: When you say that you can estimate within 3%, or, alternatively, no more than 5%, what do you mean exactly?

Expert witness: I have in mind the estimates that we surveyors make of the lowest tender that will be received back from builders when they tender for the job.

Cross-examining counsel: Now, under a *fixed fee* form there isn't a tender figure; at least, not in the normal sense. Do your figures of three or, alternatively, no more than 5% still apply in the case of a fixed fee form?

Expert witness: *(slightly taken aback)* Why, I think so.

Cross-examining counsel: *(sensing a weakness)* Do you have much experience with fixed fee forms of contract?

Expert witness: I am familiar enough.

Cross-examining counsel: How many contracts of this form have you dealt with in the past year?

Expert witness: It is not a form that I recommend to clients.

Cross-examining counsel: Will you answer the question asked, please.

Expert witness: I have not dealt with any contracts of a fixed fee form in the past year.

Cross-examining counsel: ... and in the past five years?

Expert witness: I can not recall any contract under a fixed fee form.

(pregnant silence)

Cross-examining counsel: *(new subject)* Have you studied statistics?

Expert witness: A little.

Cross-examining counsel: You did not study statistics in your course 30 years ago?

Expert witness: No.

Cross-examining counsel: You have not been to any courses on statistics as a part of your study?

Expert witness: No, I am afraid not.

Cross-examining counsel: Do you know how close surveyors get in their estimates?

Expert witness: The 3-5% that I spoke about earlier in relation to my own estimating would probably be about the level to expect.

Cross-examining counsel: You have not heard of work done going back to the 1970s on the estimating variability of surveyors?[6]

Expert witness: No.

Cross-examining counsel: Would it surprise you to hear that the estimating variability of surveyors is nearly 10%?

Expert witness: Well, it may be so. But there are so many variable factors to take account of — I do not follow how one can come up with a figure like 10%.

For example, alterations work is a very different kettle of fish from new work. One has to think of these things.

Cross-examining counsel: *(at once, picking up the opportunity afforded)* Am I right that the margin of error is likely to be greater in estimating for alteration work, as there is more uncertainty working on existing buildings than on new buildings?

Expert witness: Yes.

Cross-examining counsel: So, going back to the research in the 1970s, if the researcher had been working with existing buildings rather than new developments in the 1970s he would probably have found an estimating variability greater than 10%?

Expert witness: Perhaps so.

Cross-examining counsel: How much more do you think?

Expert witness: I cannot be sure.

Cross-examining counsel: 15%, would you think?

Expert witness: Possibly.

(pregnant silence)

Cross-examining counsel: *(new subject)* You say that the labour in the surveyor's statement of final account is excessive.

Expert witness: Yes.

Cross-examining counsel: And you have said that you took each item and estimated its cost. Did you refer to the time sheets?

Expert witness: In part, yes.

Cross-examining counsel: Only in part?

Expert witness: Yes. It was clear to me that they were excessive.

Cross-examining counsel: How did you know they were excessive?

Expert witness: Well, I priced the work.

6. *For information, see* Statistical Methods for Building Price Data *p147; D T Beeston , E&FN Spon 1983 ISBN O-419-12270-2.*

Cross-examining counsel: Did you refer to price books at all?

Expert witness: Yes. I did.

Cross-examining counsel: For what purpose did you refer to the price books?

Expert witness: Why, for help in building up a price. In other cases, I referred to the price books to check a figure from elsewhere.

Cross-examining counsel: You will have noticed no doubt that the prices in price books vary widely?

Expert witness: Yes.

Cross-examining counsel: What do you do when you find the prices differ?

Expert witness: Oh, I use the one that is the most likely.

Cross-examining counsel: You do not take an average?

Expert witness: Oh no.

Cross-examining counsel: How do you decide which of the price books to follow?

Expert witness: I make a judgment.

Cross-examining counsel: What basis do you have for that judgment?

Expert witness: Surveyors have a feel for prices, you know.

(pause)

Cross-examining counsel: *(puzzled)* Can you tell me something about that?

Expert witness: Well, I suppose not ... not explicitly.

(pause)

Cross-examining counsel: What are the constituents that you take account of in putting together a price for building work?

Expert witness: *(counting out)* I have, first, labour; then materials; then plant; and then overheads and profit.

Cross-examining counsel: What proportion of the price do you allow to labour?

Expert witness: It depends. I would expect it to be between 30% and 50%. The exact amount would vary with the type of work.

Cross-examining counsel: What proportion of the price would you expect to attach to labour in the present case?

Expert witness: *(pause for thought)* I think 45%.

Cross-examining counsel: Why?

Expert witness: Well, the work is in alterations to an existing building, so I would expect the labour content to be high — there would be little scope for high productivity. *(pause)* But again, the materials were, in general, very expensive ones. This would push up the materials as against the labour. *(pause)* So, I think 45% would be about right.

Cross-examining counsel: *(working out the arithmetic)* So the labour would be about 45% of £900,000. What does that give us? *(pause while counsel and expert witness use their calculators)*
(then bright) That's £405,000, I think. Do you agree?

Expert witness: *(beginning to feel cornered)* Yes — £405,000.

Cross-examining counsel: Now, you were saying that there was too much labour, in your opinion — to the tune of £60,000.

Expert witness: Er, yes.

Cross-examining counsel: £60,000 over a total labour cost of £405,000. *(slowly, while calculating; speaking to himself)* Now, what's that? About 15%, I think. Is that what you make it?

Expert witness: *(unhappily)* Yes, about 15%.

Cross-examining counsel: *(addressing the expert witness for verification)* The final account prepared by the surveyor allowed in just about 15% too much labour, would you say?

Expert witness: Well — er yes.

Cross-examining counsel: And there's a view, held by those whose standing in the field would be compared with your own, that the margin of uncertainty in deciding the labour content in alterations work is also some 15%?

Expert witness: Well ... yes.

Cross-examining counsel: And you have never put your own estimating to any testing that a statistician would take seriously?

Expert witness: *(involuntarily)* Oh dear!

Cross-examining counsel: I put it to you that you have no basis for believing that you can estimate to within 3% or even 5%. Is that so?

Expert witness: Well, perhaps.

Cross-examining counsel: So that you do not have any basis for believing that there is too much labour in the final account?

Expert witness: Well ... *(tailing off)* that isn't quite right, surely ...

Cross-examining counsel: *(very slightly dismissive)* Thank you.
(turning to the arbitrator) That is all I wish to ask by way of cross-examination, Sir.

Arbitrator: Very well, Mr Counsel.

Commentator: The expert witness did not fare very well under cross-examination, did he? Yet, don't you agree, he had prepared himself? ... had looked into the case carefully? ... formed what seemed to him to be a proper view? But, in truth, he was not prepared well enough.

Was he, truly, expert? Cross-examining counsel — doubtless primed by the other expert — seemed to know more than the expert did himself.

Note how important to have relevant experience. The expert had not used the fixed fee form: so how could he be expert? — a form with its own particular basis, markedly dissimilar from other forms. It is hardly surprising that counsel pounced on him. The outcome could hardly have been advantageous to the claimant's cause. Better that the expert had never taken on the job!

Could he have done better? I think he could. What do you think? Shall we give him another chance: see if he can do better? I am sure we shall — such a splendid audience! Just the start and finish of the cross-examination?

On at once!

Arbitrator: *(to cross-examining counsel)* I call on cross-examining counsel to examine the witness.

Cross-examining counsel: *(comes forward)* Thank you, Sir. *(pause)*
(to expert witness) You say you have been practising your profession of surveyor for 30 years?

Expert witness: Yes, 30 years.

Cross-examining counsel: And you qualified before that, I suppose?

Expert witness: No, I studied by correspondence course in the evenings for five years while I was working in a firm during the daytime.

Cross-examining counsel: And have you kept up with your studies at all since then?

Expert witness: Yes, I have. I have attended a number of short courses. These have been in step with my professional needs as my career has developed.

Cross-examining counsel: Have you been to any courses on estimating?

Expert witness: *(slowly, while thinking)* No, not *estimating*, precisely.
(normal speed) Estimating is a practical skill, mostly learned on the job — but there is a theoretical ...

Cross-examining counsel: *(interjects)* That's enough. I just wanted you to answer the question.

Expert witness: *(to the arbitrator)* Respectfully, Sir, I have not finished answering that question.

Arbitrator: *(to cross-examining counsel)* The witness must be given the opportunity to answer.

Cross-examining counsel: Certainly, Sir.

Arbitrator: *(to expert witness)* You were going to say? Limit yourself, please, to answering the question.

Expert witness: Thank you, Sir.
I was saying ... there is a theoretical side to estimating ... it's mostly sampling theory. I have studied this at several courses.

(slight pause)

Cross-examining counsel: How accurately can you estimate, do you think?

Expert witness: I find I am pretty much in line with what research has shown is normal — about two-thirds of the time I am within 10%.

Cross-examining counsel: *(slowly; digests)* Two thirds of the time you are within 10%. 10% of what?

Expert witness: 10% of the figure that I am predicting. In the case of work under a fixed fee form ... I mean 10% of the client's total payment to the builder (excluding later changes after the estimate).

Cross-examining counsel: How do you know that two-thirds of the time you are within 10%, as you say you are?

Expert witness: I test myself. Everyone does in my firm. It is part of our internal quality control. It is all recorded. Each individual's performance is reviewed annually.

Cross-examining counsel: *(slightly suspicious)* Do you include all your estimates?

Expert witness: Yes ... to do otherwise would invalidate the test.

Cross-examining counsel: *(trying to give a false lead)* ... and I suppose you adjust for any unusual circumstances?

Expert witness: *(firm as a rock)* No ... again, that would invalidate the test.

(pause)

Cross-examining counsel: You know that my client's expert holds the opinion that what you claim is *excessive labour* is nothing of the kind ... falls quite simply within the ordinary tolerances that apply.

Expert witness: Yes.

Cross-examining counsel: I dare say that you would agree with me ... that my client's expert is a well-qualified and experienced surveyor?

Expert witness: Yes.

Cross-examining counsel: Someone, in fact, well recognised in the industry?

Expert witness: He is a well-regarded surveyor.

Cross-examining counsel: *(guile)* Of course, it happens sometimes that two surveyors will not agree on a matter in which they are both expert?

Expert witness: Yes.

Cross-examining counsel: Each of the opinions is perfectly valid?

Expert witness: *(crucial answer)* No, that does not follow. One of the opinions may be mistaken, though honestly held.

Cross-examining counsel: *(last fling)* But in the case of estimating, in this case, in fact, the other expert's views must be quite as valid and entitled to accepting as your own views?[7]

Expert witness: The other expert's views warrant careful consideration and examination — to find if they are well founded. Upon examination, I have found them mistaken in this case.

Cross-examining counsel: *(dejected)* That is all.

Arbitrator: Very well.
(to examining counsel) Is it your wish to re-examine the witness?

Examining counsel: No thank you, Sir.

Arbitrator: *(to the expert witness)* Then it remains for me to thank you for your help in these proceedings. I think that is as far as you will need to be troubled.

Expert witness: Thank you, Sir.

Commentator: So there it is. Ask yourself these questions before accepting the job:

- Have I given my work sufficient thought?
- Have I taken pains to keep up-to-date?
- Am I familiar, on a day-to-day basis, with the particular context of the dispute?

In fact, am I, truly, an expert here?

(deliberately) Then, once having decided to take the job on, bear in mind that *the other side's position may have much more to be said for it than appears at first.*

End

7. He who affirms must prove. For proof there must be a positive tilting of the balance in favour of what is being affirmed. Here we have the claimant affirming that there is too much labour in the final account. For proof of this the claimant is relying on the opinion of the expert. If cross-examining counsel can persuade the expert to agree that his opinion is no better than that of the other expert, then the claimant will not have tilted the balance: will not have proved his point.

Key data

The surveyor as expert witness: Building and development play

A Bruce Shaw play by Alan Morris

Final account

Final account for works carried out at:
39 The Oaks, Nuneaton

All as estimate of 23 September 2004	£500,000
Additional items:	
● Renew front door	
● Take down plaster and renew — kitchen	
● Bricking up window opening — study	
● Renewal of felt roofing — porch	
● Redecoration — dining room	
Total additional items	£400,000
Total	**£900,000**

Claimant's expert's reworking of the final account

Final account for works carried out at:
39 The Oaks, Nuneaton

Works originally ordered	£680,000
Additional items	£160,000
Total	**£840,000**
(£60,000 overpaid)	